Samantha's Search

3D Shapes

Kathleen L. Stone

Dedication

Many of my family members and friends are wonderful gardeners. I envy them their skills. Someday I hope to have a beautiful garden … maybe when I retire! In the meantime I will enjoy the beauty of their gardens vicariously. I dedicate this book to gardeners (and wannabe gardeners) everywhere!

In a lovely little garden
So colorful and merry
You might catch a glimpse
Of Samantha the Fairy.

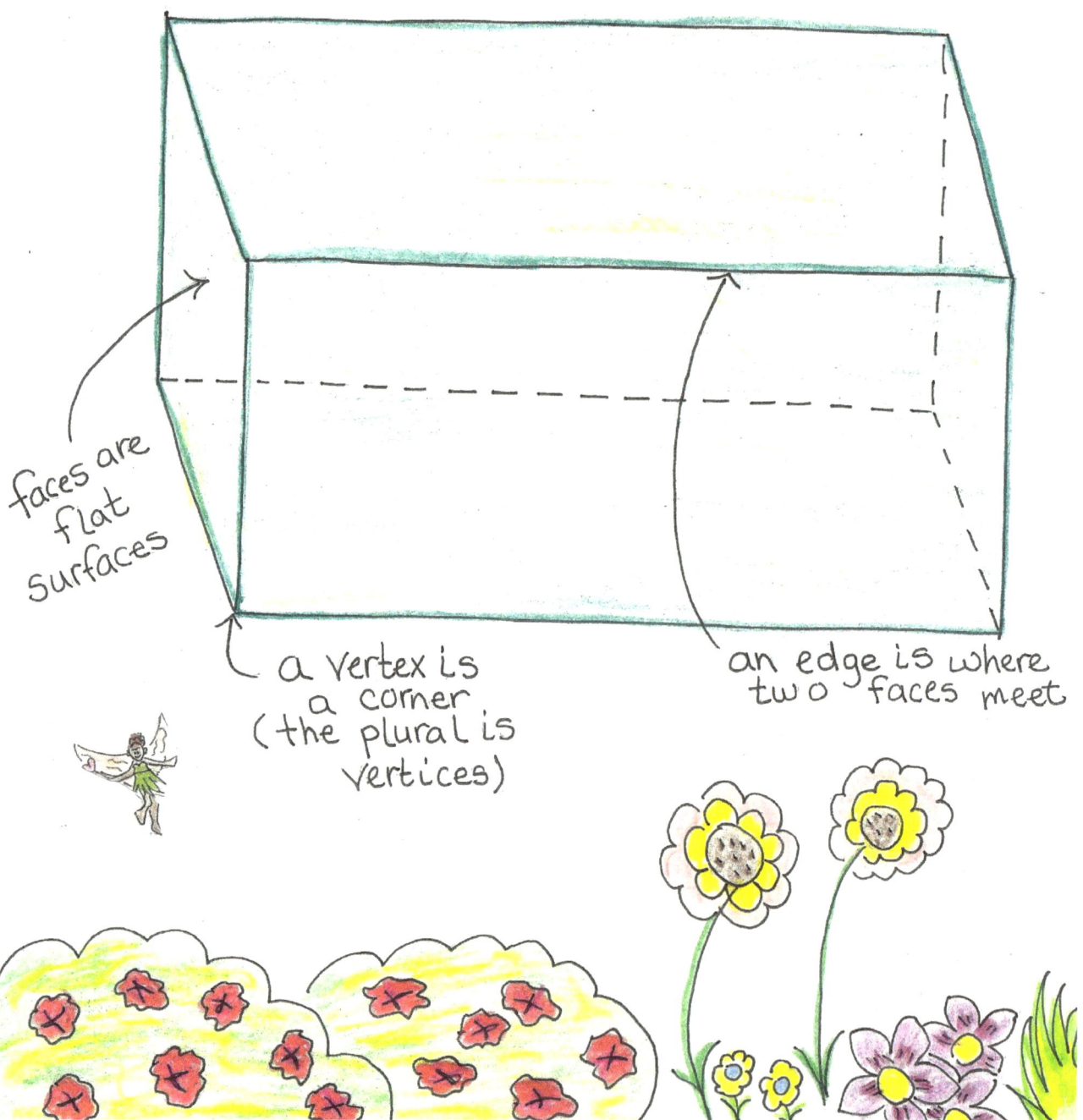

faces are flat surfaces

a vertex is a corner (the plural is vertices)

an edge is where two faces meet

Today Samantha is flying
Here and there
In search of 3D shapes
She's looking everywhere.

3D shapes are solid.
They are not flat.
They have surfaces, edges, and vertices.
Did you already know that?

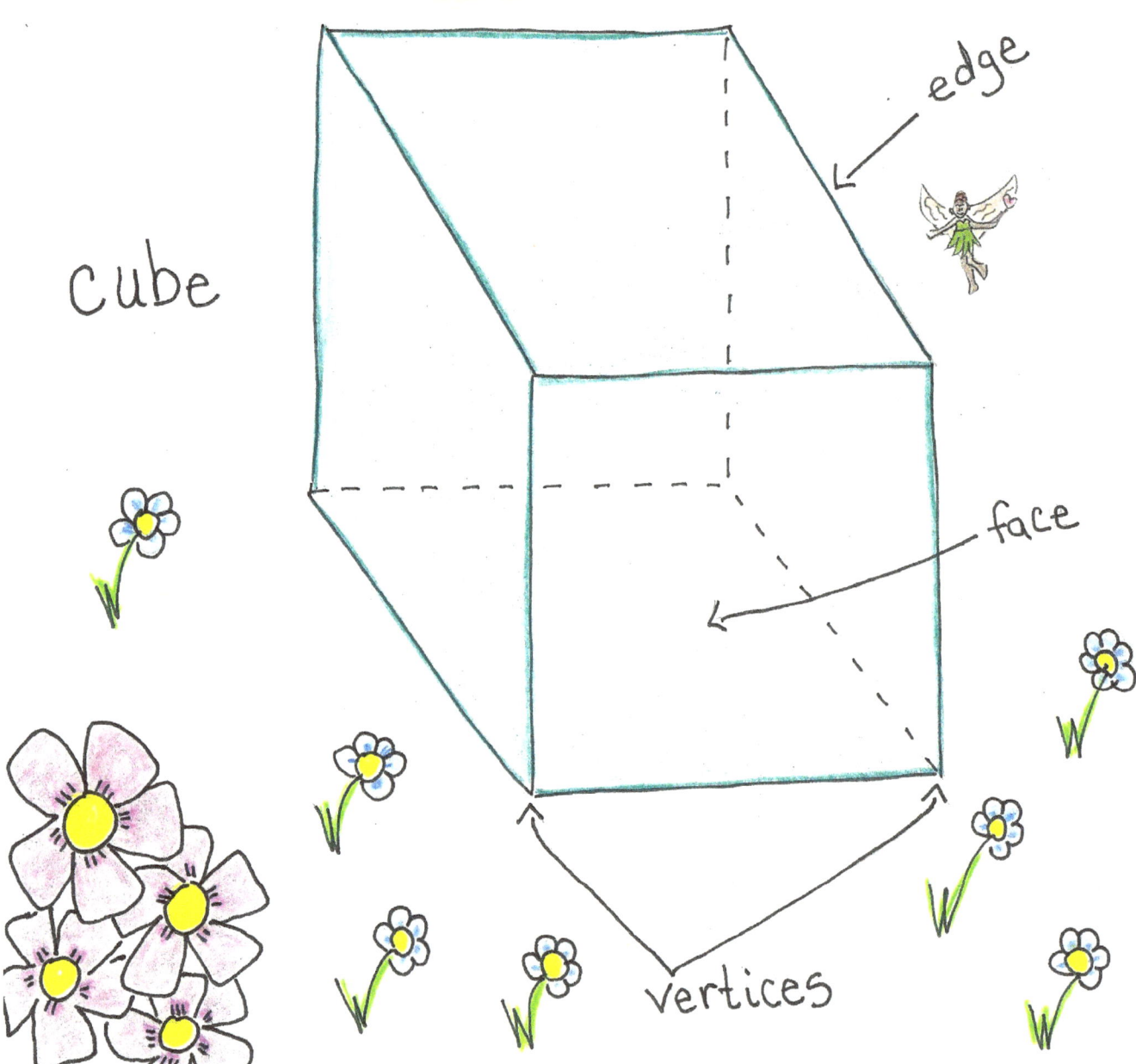

cube

edge

face

vertices

A **cube** is the first shape
Samantha is looking for.
Let's tag along with her
As she sets off on her chore.

Cubes have *six* faces,
Eight vertices, and *twelve* edges.
Samantha sees several **cubes**
Over there by the hedges.

There's a pair of dice and
A pretty purple box,
An ice cube, sugar cubes,
And some cute baby blocks.

cone

vertex

face

A **cone** is the next shape
Samantha hopes to find.
We better hurry up or
She'll leave us behind.

Cones have a curved surface,
A vertex and *one* flat face.
Samantha spies several **cones**
Next to that beautiful Queen Anne's
Lace.

Samantha sees a party hat,
A funnel and traffic cone.
She also sees ice cream
And a polka-dot megaphone.

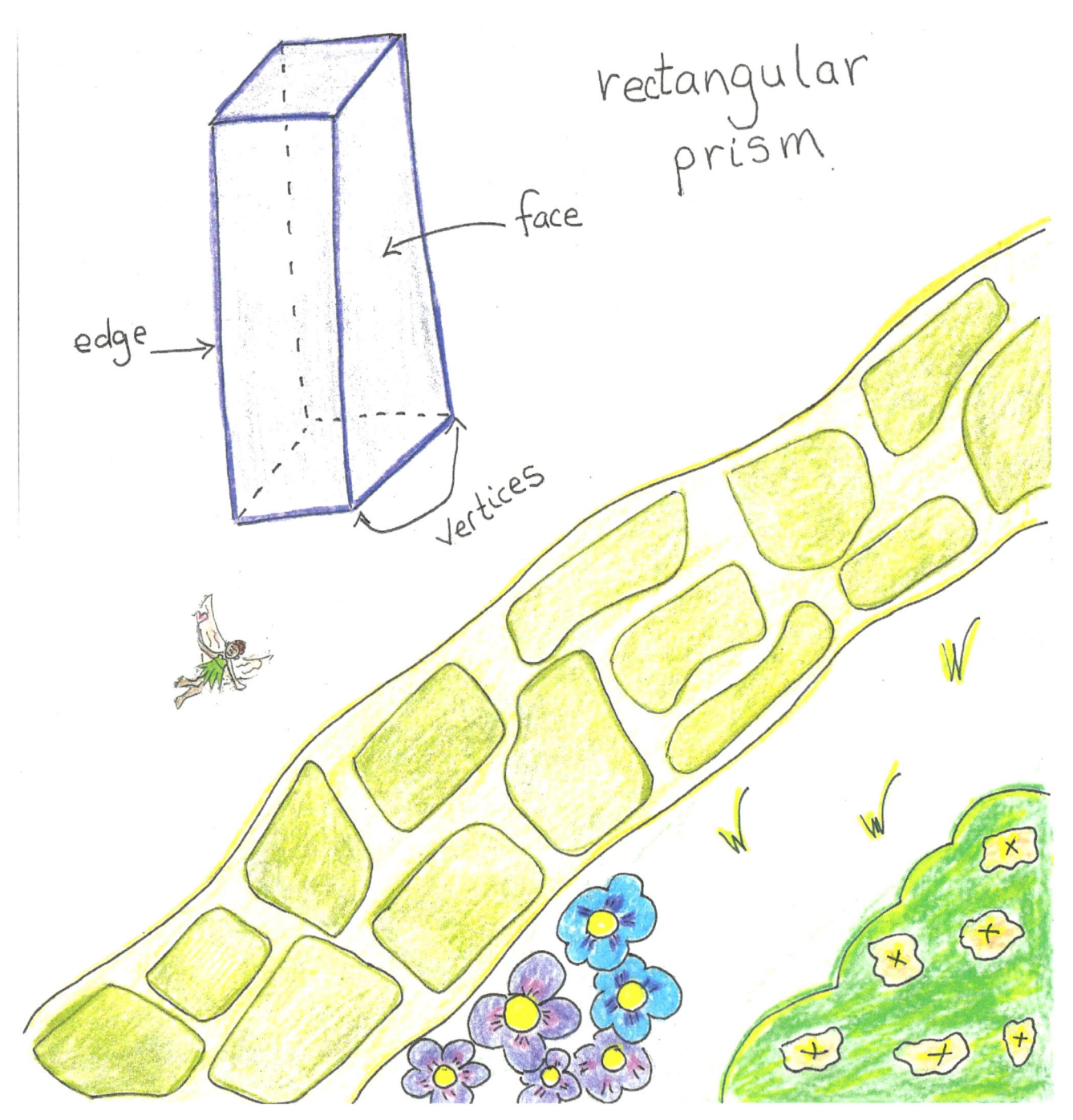

rectangular prism

face

edge

vertices

Samantha hopes to find
A **rectangular prism** next.
She looks over by the flowers
And finds just what she expects.

This shape also has *eight* vertices,
Twelve edges, and *six* faces.
You can find **rectangular prisms**
In all kinds of places.

Do you see the tissues
And the yellow cereal box?
There's a deck of cards, the back of a
truck,
And a pretty pink jewelry box.

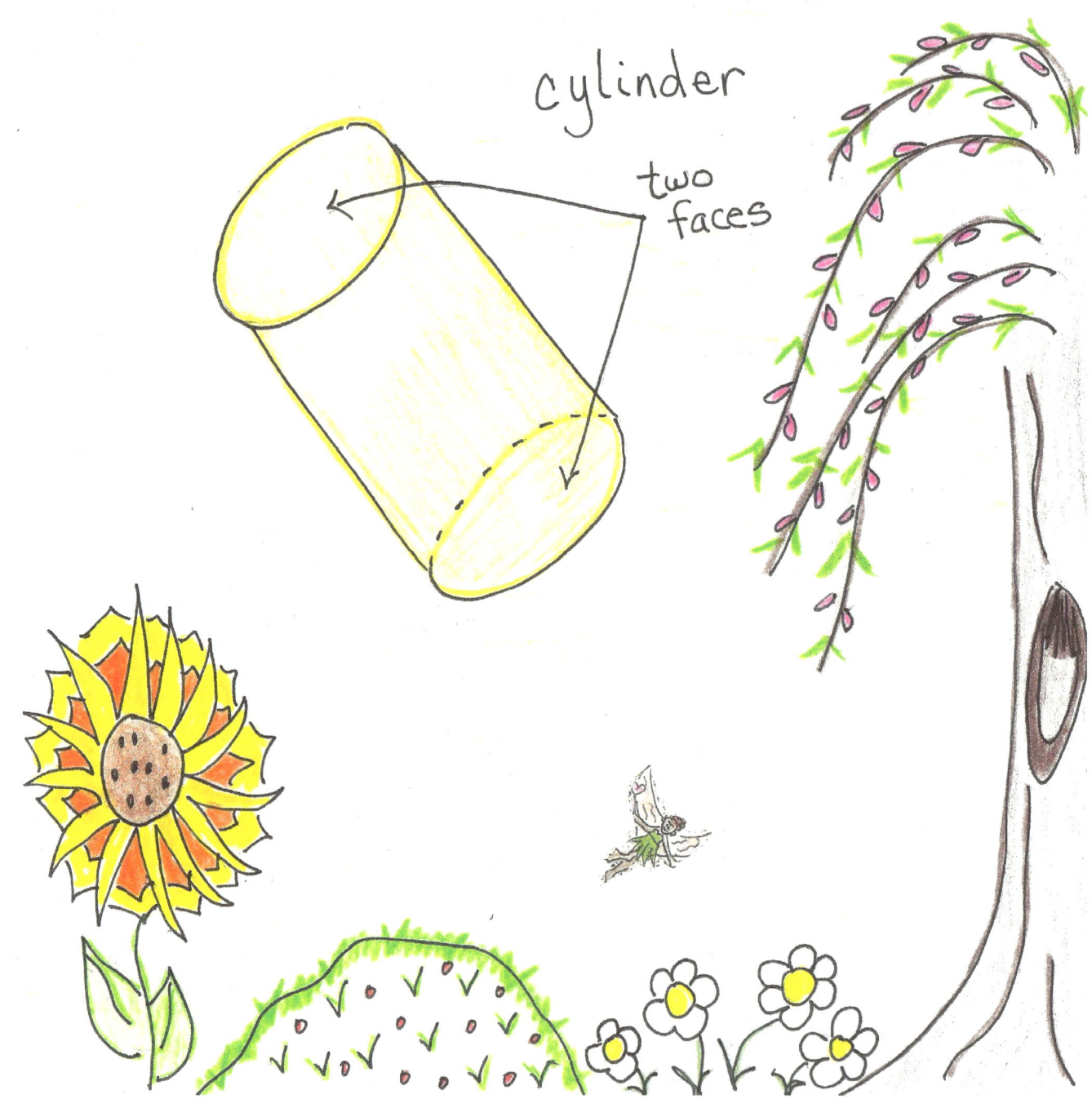

cylinder

two
faces

Samantha is off again
Looking for more shapes.
She thinks she see some **cylinders**
Near those luscious looking grapes.

Cylinders have a curved surface
And *two* flat faces.
Samantha counts *five* **cylinders**
Hiding in all those spaces.

There's a smelly garbage can,
And a big bass drum.
Some tuna and soup,
And a can of yummy plums.

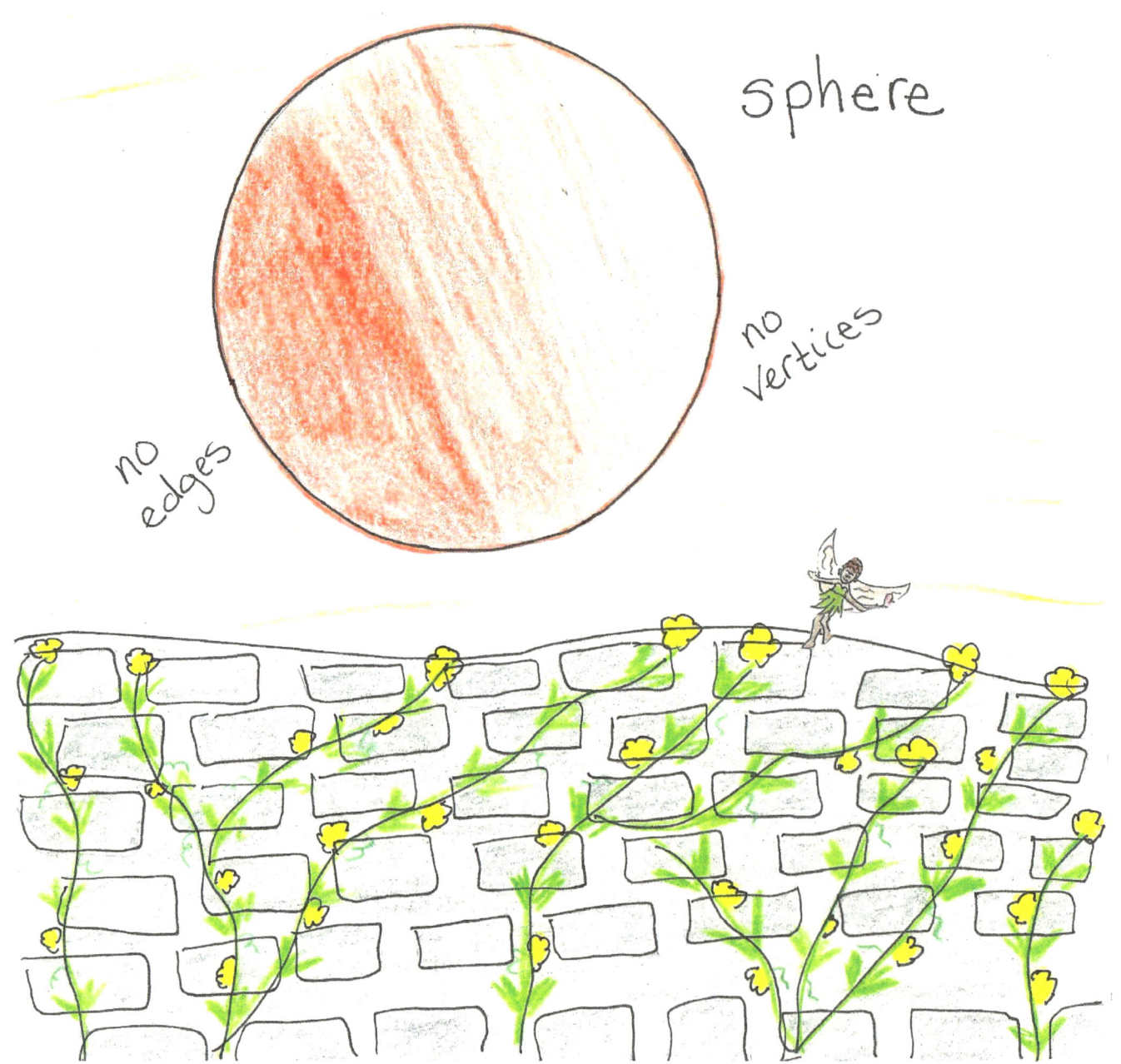

sphere

no vertices

no edges

There's a very special shape
That Samantha has in mind.
She's confident a **sphere**
Will be easy to find.

Spheres are perfectly symmetrical
With no edges or vertices.
Samantha finds several **spheres**
Over there by the trees.

Grapes and oranges
And all those different balls.
But the marbles are Samantha's
Favorite **spheres** of all.

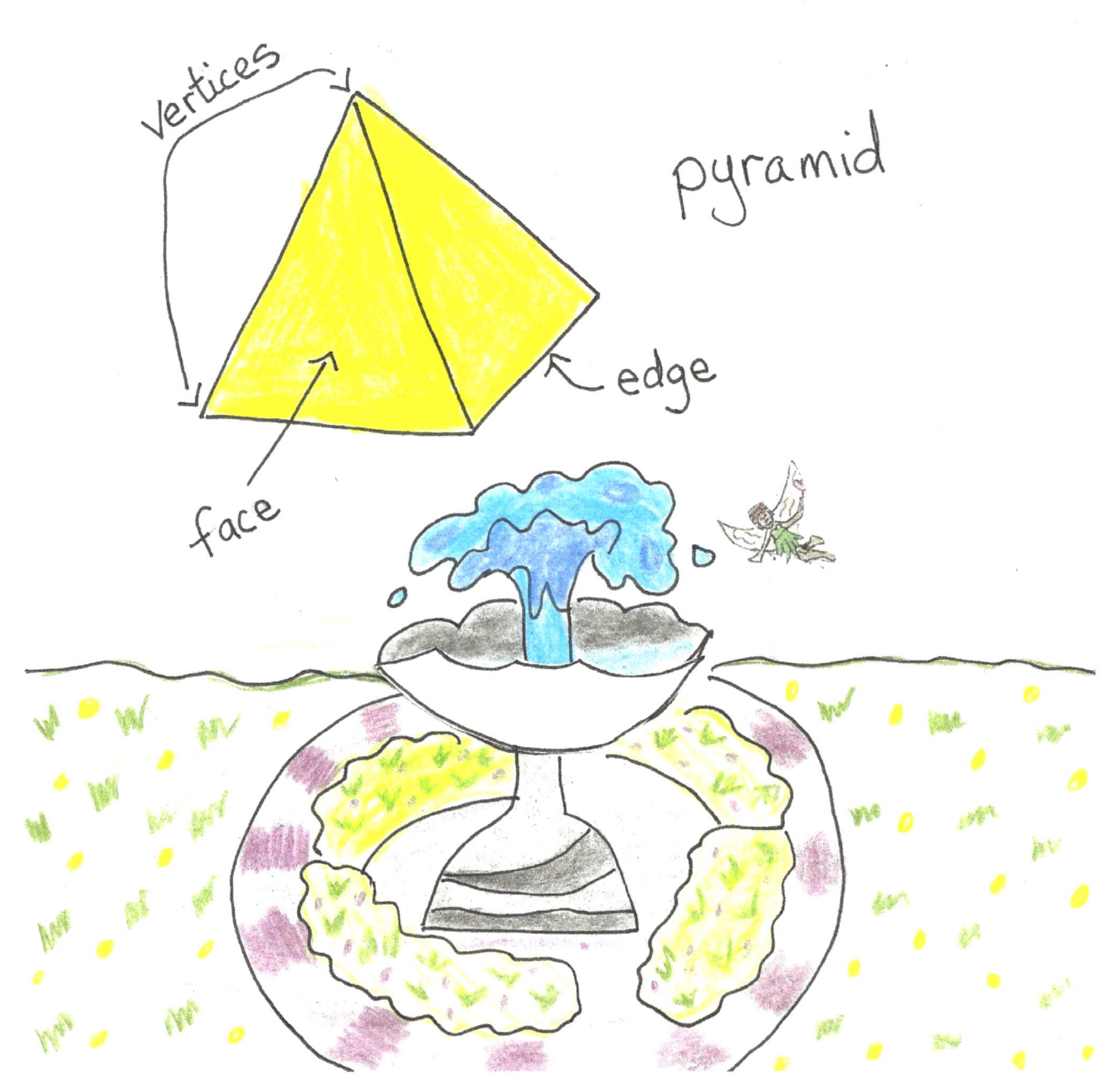

vertices

pyramid

edge

face

Samantha is afraid
A **pyramid** she won't spot.
She is about to give up
Then she sees a flower pot.

Now **pyramids** have *five* faces,
Eight edges, and *five* vertices.
Samantha spies *three* **pyramids**
With vines blowing softly in the breeze.

Samantha has all her shapes
And is headed back to her spot.
Finding all these 3D shapes
Was much easier than she thought.

With a lot of hard work
And some fairy dust too,
Samantha used all her shapes
And created something new.

3D Shapes

Learning about 3D shapes is a great opportunity for a variety of hands-on activities. A great place to begin is by looking at real life examples of 3D shapes. As you discuss 3D shapes and their attributes it is important to use the actual names of each shape (i.e. *cube* rather than *box*) as well as other key vocabulary words including *surface, curved, flat, face, corner, vertex, vertices, and edge.*

Enrichment Activities

Shape Sort

Materials needed:

Plastic or wooden 3D shape models
variety of pictures or actual "real life" objects that might include:
> **cube** … Rubix cube, sugar cube, dice, alphabet block, etc.
> **rectangular prism** … tissue box, deck of cards, juice box, supply box, etc.
> **cone** … funnel, mini megaphone, party hat, mini traffic cone, etc.
> **cylinder** … variety of cans (soda pop, fruit, tuna, etc.), toilet paper roll, candle,
> **sphere** … marbles, variety of balls (tennis, golf, ping pong, etc.), tangerine, etc.

Have the children use the pictures and/or objects to sort by various attributes:
- 3D shape (cubes, cylinders, spheres, etc.)
- shapes with vertices and shapes without vertices
- shapes with flat surfaces and shapes without flat surfaces
- shapes that can roll, slide, or stack
- shapes with curved surfaces and shapes without curved surfaces … this might be a nice time to introduce a new vocab word, *polyhedron* (a shape with no curved surfaces)

Challenge Activity … place the "real life" objects in a bag and have the children "blindly" identify each shape by "feel" only.

Food Sort

This activity is similar to the activity **Shape Sort** but uses food items that might include:

> **rectangular prism** … Starburst candy, Rice Krispies treats, juice box or wafer cookies
> **cube** … cheese cube, Jell-O Jiggler cube, or caramel cube
> **cone** … Bugles, Hersey Kiss, or ice cream cone
> **cylinder** … marshmallow (medium and small)
> **sphere** …grapes, Whoppers, cheese puff balls, Kix (or other spherical cereal), or an orange

Build a Shape

Materials needed:

modeling clay
toothpicks

   OR

pretzel sticks
mini marshmallows
napkins

Use clay and toothpicks to make *rectangular prisms, cubes, and pyramids*. To add a little challenge (and fun) use pretzel sticks and mini marshmallows. The pretzel sticks become the *edges* and the mini marshmallows the *vertices*.

ABOUT THE AUTHOR

This is Kathleen Stone's fourteenth book. Kathleen combines her love of math, literacy, and teaching in all her books. Not only does she use her books in her own second grade classroom, but many other teachers have contacted her to say how much they and their students enjoy her books (including several Middle School Special Education math teachers who tell her that their students enjoy how she makes math more meaningful for them).

Kathleen is a National Board Certified educator, with over thirty-five years teaching experience, and teaches second grade with North Thurston Public Schools. She is also a volunteer with West Thurston Regional Fire, where she and her husband work as Juvenile Firesetter Interventionists (working with children and their families that have been involved in fire setting situations). When not teaching, she can often be found reading a good mystery by the lake or spending time with her husband, sons, daughter-in-law, and grandchildren.

Math is all around us
No matter where you turn
Open your mind to the wonders of math
And all that you can learn

www.ingramcontent.com/pod-product-compliance
Lightning Source LLC
Chambersburg PA
CBHW050358180526
45159CB00005B/2065